In summertime you might spot these insects. They are fuzzy buzzers. They are pretty and they are speedy. They are bumblebees!

A bumblebee has pretty yellow and black fuzz. Small wings fly the bee's big chubby body. It is funny to see it up close.

Queen bees fly around on sunny mornings in spring. They find holes in the dirt for nests.

Here is a handy chipmunk hole. It will do just fine for this queen bee, thank you!

The queen makes a nifty little pot. She fills the pot with sweet, sticky nectar. Then she sits on her eggs like a hen! She sips from the pot if she gets hungry.

Soon there are lots of bees. They fly out of the nest. Then they look for pretty blooms. Their sturdy wings flap too fast for us to see!

Bumblebees can sting. But they are not grumpy. They won't bother you. They just want to find yummy pollen.

The bees get pollen from lots of plants. A poppy! A pansy! A cherry blossom! Bees are not fussy. They might fly around gardens, or grasslands, or farms.

The bees carry pollen to the nest.
One nest can have hundreds of bees!
All the bees come from the eggs of just
one queen. Isn't that something!